TRACTOR HERITAGE

CASE
CASE

TRACTOR HERITAGE

Photography by Duncan Wherrett

Text by Trevor Innes

First Published in Great Britain in 1994 by Osprey, an imprint of Reed Books Limited, Michelin House, 81 Fulham Road, London SW3 6RB and Auckland, Melbourne, Singapore and Toronto

Reprinted Autumn 1995

ISBN 1 85532 411 3

Editor Shaun Barrington
Page Design by Paul Kime/Ward Peacock Partnership
Printed in Hong Kong

Half title page
1910 Caldwell Vale; (see page 23). The power steering arrangement can be seen in the lower part of the picture. It was worked by engaging one of two cone clutches and the engaged cogs would wind along a screw cut into the tie rod steering linkage. With the power steering, it was claimed a 6 year old boy could drive it. All this in 1910. Although costing £1250 (10 years wages at the time), many were used in the construction of the newly-formed Federation of Australia's new capital city – Canberra. The failure of the company in 1916 seems to have been due to over expansion, as in 1914 they were also offering a 40-50 hp farm tractor, stationary engines, and had developed a four wheel drive, four wheel steer car. Of the 32 made originally, the one here is the only survivor of the steel wheeled version. Another two incomplete examples are still in existence

Title page
1936 Case Model C, one of several model variants; (see page 27)

Front cover
1950 Turner 4V95, built in Wolverhampton, England; (see page 126)

Back cover
The Caterpillar Tractor Company's 1927 Holt 2 Ton; (see page 61)

This book is dedicated to the memory of John McCallum, photographed here on pages 10 and 20 driving his beloved Benz. He was one of the founder members of the Booleroo Steam and Tractor Preservation Society, and his ideas and dedication will be sorely missed.

Acknowledgements

We would like to thank the following people for their help in putting this book together, especially the moving of over 200 tons of tractors and machinery around the countryside, which was not an easy task. Daniel Allen; Devon Amber; Ken Arthur; David Ball – Bluey's Restaurant, Melrose; Brian Beyer; Birdwood Motor Museum; Booleroo Agencies Pty Ltd; Lynne Brandon; Dave Butt; Ian Catford; Roger Chapman; David Correll; Tom Correll; Mike Hamish; Vern Harvie; Ferg Innes; Gwen Innes; Neil Innes; Wallace Jolly; Les and Bob Jones; Kadina Museum; Hedley Koch; D. Kotz; Brian Knauerhase; Ian Kumnick; Ernie Masters; John McCallum; Norm McCallum; The Mount Remarkable District Council; Brian Neumann; Geoff Nicholls; Jim Nicholls; David Piggott; Ron Piggott; Lester Reichstein; Merv Robinson; Kev Rohrlach Museum, Tununda; Rob Schmidt; Don Wilsdon and the Geralka Rural Farm, Spalding.

Overleaf

This group photograph illustrates the three main developments of the tractor. The first, the Fowler single cylinder ploughing engine, was delivered new to a pastoral company in South Australia in 1876, and is believed to be one of the oldest self-propelled ploughing engines in existence. It weighs around 20 tons and required most of its power to propel itself along the ground. Used for sinking dams, all of the work was done with its winch mounted under the boiler. It was last used to drive a shearing plant in the early 1940s – a useful life of some 70 years.The second machine, the International Mogul 10-20, shows the first developments of what we now regard as the modern tractor. It had a single cylinder four stroke gasoline stationary engine producing 20 hp at 240 rpm.The third tractor is a Minneapolis Moline made around 1938. It shows another big step in tractor development – the pneumatic tractor tyre. This improved the drawbar hp by as much as 25% in many cases and thus secured the basic tractor format which is still in use today. Tractors could then be used on well made roads, whereas previously the lugs on the steel wheels tore the roads apart.The Clutterbuck is a stationary engine made in Australia, and is similar to many hundreds of different makes, of which only a few are left today. Because these engines were on wheels, they could be used for numerous farm operations, from driving a winnower to pumping water

BOOLEROO ST
PRESERVATI
CLUTTE
OIL EN

Contents

Foreword by Trevor Innes

My first recollection of tractors was of riding on a specially made seat on a WD9 International tractor in the early 1960s. My father, Ferg Innes, was always interested in things mechanical and started collecting vintage tractors in the mid 1960s. The re-creation for this book of the period working scenes from the 1920s held a particular fascination for me, because I realized that even my father only started farming in the late 1930s and had never seen the early tractors working. Everyone involved learned a great deal; even though the tractors are used each year for the Show Days, they are never used in actual working conditions.

The Booleroo Steam and Tractor Preservation Society was formed in Booleroo Centre in 1968 by a group of local farmers, with a common interest in the preservation of farm machinery, showing the progress of farm mechanization. The aim was to have as wide a range of machines as possible, covering tractors, stationary engines, steam engines and farm implements. A machine has to be in a restored working condition before it is allowed into the collection. With a great deal of work now having been done, the Society receives visitors from all over the world. The collection is in Booleroo Centre, a small town surrounded by farm land, about 160 miles north of Adelaide, South Australia. Most of the machines pictured in this book are from this collection and the remainder were all within a 150 mile radius. Apart from the enthusiasm and efforts of the Society members, there are two main reasons why so many rare machines exist in this region: firstly, as they became redundant and were left in a field or barn, they deteriorated little in the dry climate of South Australia; and secondly, they were not melted down as scrap metal during World War 2, as happened in some countries.

The stories included in this book will, I hope, give an insight into some of the problems associated with early mechanized farming. Many surprises have come to light during research and every effort has been made to be as accurate as possible, although often there is very little information on some of the earlier tractors.

The 1926 Cross Engine Case 18-32; *(see page 25)*

Above
Pastoral scene featuring a 1927 John Deere Model D, the first tractor from the famous company

Right
A 1922 Peterborough L30.4i; an engine full of good intentions that were not always followed through; *(see page 121)*

The Steel Horse

Before the advent of the internal combustion engine, steam power by means of traction engines was playing a minor but increasing role in providing mechanical muscle on the farm. Such machines had certain disadvantages, however, and this restricted their usage. For one thing they were very expensive. They were also big, heavy and inefficient and they were mainly used for winch power. The uses to which they could be put were somewhat limited. The 1875 Fowler ploughing engine was a typical example. Weighing 20 tons, it required ten people to operate it. The engine pulled two water carts, and two horse drawn wagons continuously worked to feed the machine with wood and water. The Fowler ploughing engine was finally stranded with insufficient supplies of either, and had to be towed home by 40 camels! That proved to be its last outing. In a similar vein, a stationary steam engine at Tarcoola in South Australia used up all of the wood within a 35-mile radius over its 50-year working life. Towards the end of the 19th century, the internal combustion engine was being rapidly developed, and a vast range of stationary engines started to appear. These began to be used for such functions as driving threshing machines and pumping water, and due to their relatively small size and weight it was a simple matter to put them on a wheeled trolley in order to move them to different places on the farm. From this stage, it was a natural progression for the wheels to be driven by the engine itself and thus the first tractors were on their way. From this time, traction engines, stationary engines and the new tractors continued to develop along their own lines.

Left and overleaf

For nearly 40 years there raged the controversy – which was better – the horse or the newfangled steel horse – the smelly, noisy tractor? Now that tractors have completely replaced them it is easy to underestimate just how powerful horses were and the enormous role they played in agriculture throughout the centuries. By way of comparison, the Benz Sendling is matched here against a team of eight Clydesdale horses, both pulling similar ploughs. The Benz engine developed 34hp, but losses through the transmission and the steel wheels left only 17hp at the drawbar. The rule of thumb is that only 75% of maximum load should be pulled continuously in order to avoid mechanical failure, which leaves 12.75 'rated' hp. Each mighty Clydesdale was considered to be worth 4 rated hp. (The original 'hp' was the equivalent of the average pit pony down a mine, thus a horse became equal to 2.5hp.) Best estimate? The eight horse team here can pull well over twice the load of the tractor, despite the paper match-up of a 30-34hp machine and 32hp of Clydesdale muscle.

1925 ALLIS-CHALMERS 18-30. *Made by the Allis-Chalmers Co. in Milwaukee, Wisconsin, U.S.A., this model was in production for ten years from 1919 to 1929. Over this period, virtually none was made in 1922 due to the tremendous marketing of the Fordson and the post-war depression. Peak production of 4760 was reached in 1928, and a total of 16,000 were built in all. The engine, with its bore of $4\frac{3}{4}$" and stroke of $6\frac{1}{2}$", produced 37 rated belt horsepower and 19 rated drawbar horsepower at 930 rpm. There were two forward speeds of 2.58 and 3.16 mph with a reverse of $3\frac{1}{4}$ mph and it weighed 6640 lbs. The engine was an excellent unit and was used in numerous other applications. The tractor used 108 gallons of kerosene to plough 120 acres. The testimonial from one owner was; "If I were asked to compare the Allis- Chalmers to a team of 10 – 20 horses, I would say she stands in the same relationship as a motor car to a buggy and pair"*

1916 BATES STEEL MULE. *The Steel Mule was made by the Bates Machine & Tractor Co. in Joilet, Illinois, U.S.A. It had a 4 cylinder 30 hp (13 drawbar hp) engine running at 900 rpm. The horizontal tubular radiator was made along the lines of a steam boiler, because cooling systems had not yet been fully developed. The Bates, however, had a water pump and fan, which was at the forefront of technology at the time. A Dixie magneto with impulse started, Bennett carburettor, with air cleaner, and a Pickering governor were also fitted. The Bates had a single track drive at 2¼ – 3½ mph. It weighed 2½ tons and cost $865 when new. Why buy it? The sales brochure at the time listed its many accomplishments: it could plough 10 acres/day, cultivate 31 acres/day, harvest 40 acres/day, and mow 36 acres/day. (It didn't specify what kind of acres!)*

Although the whole machine could only be described as bizarre, perhaps its most unusual feature was its three driver's wheels. These were stacked on top of one another in decreasing diameters, the first being the steering wheel, the second the clutch and the third the throttle. This actually makes more sense than at first sight, as they were designed to fit a shaft in the middle. With extensions fitted, the tractor could be controlled from behind with the driver sitting on the attached implement. In this way, one person could drive the tractor and adjust the attachment at the same time

BATES STEEL MULE

1929 BATES STEEL MULE. *The Bates Model F 18-25 was in production for 16 years from 1921 to 1937. It was a half-track machine with steering by wheels at the front instead of clutches or brakes as on full-track crawler tractors. Its weight was 4850 lbs, and it had a belt pulley on top of the gearbox. Much of the life of this machine was spent building dam banks with a scoop along the Murray River (the largest river in Australia) to prevent flooding, and doing belt pulley work driving irrigation pumps. It is seen here pulling a road grader, which is a machine for levelling and grading a non-sealed road, that is, to form a camber to run the water to the edges. The suspended strip can be adjusted to move dirt or gravel in many different directions and at different angles. The Bates Model F Steel Mule had a Le-Roi 4 cylinder side-valve engine with a bore of 4¼" and a stroke of 6", developing 25 belt hp and 18 drawbar hp. It had a honeycomb radiator, a Kingston carburettor and a 10 gallon fuel capacity. The steering box was of poor design and a weak point, as it had a tendency to wrench itself to full lock, which certainly helped to make life a little more exciting. This is one of the numerous machines restored by Lynne Brandon*

1927 BENZ SENDLING. *Benz-Sendling Motorpfluge of Berlin, Germany, manufactured the Benz Sendling in 1927. It has perhaps the most technically advanced engine of its time, being the world's first commercially produced full fuel injection engine. Unfortunately, the rest of the machine was rather crude, especially compared with some of the tractors being made in the U.S.A. at the time. It had a two cylinder, 5⅜" bore, overhead valve diesel engine, which developed 30-34 hp at 800 rpm. There was a one gear, forward and reverse gearbox driving through a single rear wheel. The Benz used about 1½ gallons of fuel an hour, and required a full set of muscles to steer it, especially under load. As they were very difficult to start, even on half compression, many people simply left them running 24 hours a day during busy periods. One of the locals devised a method of turning his engine off next to a stationary engine and then starting the Benz through a belt drive the next day.*

This particular tractor was bought during the depression, and when the payments could not be maintained the agents removed the cylinder heads, so that it could not be used. Now that the tractor could no longer earn its keep, it was not clear how the owner was supposed to pay for it. Consequently it simply fell into a state of disrepair and was abandoned. It was restored and is driven here by John McCallum

1950 BROCKHOUSE PRESIDENT.
The President was made by Brockhouse Engineering Southport Ltd, in Lancashire, England. This beautifully restored example cost £1500 in 1950, with a rotary hoe cultivator included. The engine is a four cylinder Morris 8 car engine of 918 cc, rated at 27.6 bhp at 4400 rpm, from a 2¼" bore and a 3½" stroke. A wide range of accessories was provided, from a mid-lift linkage system featuring a double-bar tool frame for hoeing, ridging, etc., to a 3 point linkage worked off a gear pump, which slipped into the side of the transmission and could lift 5 cwt. at 500 psi working pressure. The tractor was primarily designed for orchard, market garden and odd-job use and could have the optional lighting kit fitted

CASE

Above

1910 CALDWELL VALE. *This machine was a technological marvel in its time, having four wheel drive, limited slip differentials, power steering, and a square bore to stroke ratio, which was unheard of at the time. Designed by two Adelaide men, Felix and Norman Caldwell, it was made in Auburn, New South Wales, Australia. It had a four cylinder, T head, side-valve engine of 6½" bore and stroke, delivering 80 hp at 800 rpm. There was a twin ignition system – 6 volt trembler coils for starting and a Simms magneto when running, and the carburettor was a 3½" Schebler. With a fuel consumption of 1 mpg, a 60 gallon fuel tank was necessary. The valves were a massive 3⅞" in diameter. The 1912 models were fitted with solid rubber tyres and three-way tipping bodies. This example has specially designed sand wheels 5 ft in diameter and about 16" wide, and the normal operating speed was 5 mph with a load of 25 tons, although it could haul up to 45 tons with a number of trailers. Its success was such that with almost non-existent roads, even when loaded with 25 tons, it was said that the only thing that would stop it was an inch of rain*

Left

1919 CASE 15-27. *The Case 15-27 was made by the J.I. Case Threshing Machine Co., in Racine, Wisconsin, U.S.A., and it was the second of the cast frame tractors. Its four vertical cross-engined cylinders were cast en bloc. The bore was 4½" and the stroke 6", putting out 27 rated belt hp at 900rpm. The transmission was two speed 2¼ and 3 mph, with reverse. This model weighed 6460 lbs and could pull 2850 lbs in low gear. The price was $1800 in 1920 and in 1925 it was superseded by the 18-32 model*

CASE

Above

1926 CROSS ENGINE CASE 18-32. *This model was uprated from the 15-27 in 1924 to give 5 belt hp more and 3 drawbar hp. They remained in production until 1928, when the new, lighter in-line engines came out. The specifications are much the same as for the 15-27, but the engine ran at 1000 rather than 900rpm, and different gear ratios were fitted, namely 2.46 and 3.48 mph. Weight was 6680 lbs. The cross engined Cases established the company as one of the world's foremost tractor producers, with a reputation for reliability and longevity. In its heyday, this tractor farmed up to 2000 acres a year, which adds up to a lot of hours in the saddle. The above tractor still starts first time and runs like a Swiss watch, which is a credit to both the engineers who designed it and to the man who restored it, Hedley Koch. An old story is supposed to be based around one of these machines: a father, who hated tractors and the thought of his horses being superseded, went to take over from his son who had left the tractor idling during lunch. As it was running on kerosene, it had cooled off and was spluttering. The son called out; "Choke 'er, Dad! Choke 'er!" "Choke 'er. I'll bloody well throttle 'er!", came the exasperated reply*

Left

1927 CROSS ENGINE CASE 12-20. *This little tractor first appeared in 1922 and remained in production until 1928, producing 21 rated belt hp and 13 rated drawbar hp. The engine had a* $4\frac{1}{8}$" *bore and a* 5" *stroke, and ran at 1050 rpm, with two forward gears of 2.02 and 3 mph. It was a cast frame tractor weighing 4450 lbs, yet was still able to pull 3150 lbs. Its reliability was such that this particular example was working in a wood yard in 1971, having been in use for 44 years*

1936 CASE C. *The Case model C updated the somewhat old fashioned and unconventional cross engined models, having an in-line four cylinder OHV petrol/kero engine with a 3⅞" bore and a 5½" stroke. It produced 25 rated belt hp and 16 rated drawbar hp at 1100 rpm. The Case C had 3 forward gears up to 4.5 mph, and one reverse gear. Several different models were on offer: the CC being the row crop model; the CO a special tractor designed for orchards; and the CI, which was an industrial usage tractor. A power take-off shaft was available on all C models, as an option. An interesting testimonial from the time reads as follows: "I am very pleased with our Case model CC tractor. I like the independent brakes to make square corners when ploughing. I also like the open wheels because it is nearly impossible to fill the lugs. I have done nearly all the field work on our 160 acre farm, although I am only 13 years old." Marion Bergren, Red Oak, Iowa*

TWENTY

Left

1928 CATERPILLAR 20. *The Caterpillar 20 is set in a blacksmith's forge built in 1865 in Melrose. The tractor was made in Peoria, Illinois, U.S.A., by the Caterpillar Tractor Co. It weighed 7822 lbs and sold for $1900. The OHV engine of 4" bore and 5½" stroke produced 26 rated belt hp and 20 rated drawbar hp at 1100rpm. It had three forward gears giving a maximum of nearly 5 mph, and a reverse gear. A Caterpillar 20 was ploughing a field and, while going over an old water-course, started to sink. Immediately the plough was unhooked, but the tractor continued to sink until it was buried to over the top of the tracks. The farmer's son was sent to fetch some railway sleepers, which were bolted across the tracks to form its own road, but when started the tractor just drove the wooden track down into the bog. It then took a day's work with a Caterpillar D4 to pull it out*

Above

1934 CATERPILLAR TWENTY TWO. *This was one of the first of the yellow and black, caterpillar-coloured tractors, which became the standard colours after the merger of the Best and Holt companies in 1925. The Caterpillar 22 was basically an up-rated 20. It ran on petrol, with about one hp less when running on distillate. One of the local farmers had two Twenty Two's and a Two Ton, which were worked around the field together. To save on manpower before power take-off and hydraulics were fitted, farmers would sit on the implements and drive the tractor through rope reins, or an extension steering wheel, to control the steering, and with a system of pulleys to control the throttle and clutch. One day, while driving the Twenty Two in this way, the yoke on the seed drill broke and the tractor took off on its own. The farmer chased after it and jumping on managed to shut it off, only to look down and find that his foot was not on the step but in between the drive sprocket and the caterpillar track. Backing it off, his boot was not even marked, but his toe had been completely cut off inside the boot. After being rushed to hospital, it became gangrenous and his leg was going to be taken off at the knee. Luckily they had just received stocks of a new drug called 'penicillin', which miraculously saved his leg*

Above
1954 CATERPILLAR D2 4U. *The D2 was the beginning of a new line of diesel Caterpillars which were started by a small petrol engine. The main engine was four cycle and compression ignition, with four cylinders and 4" bore and 5" stroke. 32 rated belt hp and 24 rated drawbar hp were produced at 1525 rpm. The maximum drawbar pull of 6,778 lbs was considerably more than equivalently horse-powered wheel tractors of the time. Five forward gears were provided and it could turn in a radius of 10 feet. The D2 entered production in 1937, with four models produced until 1958, by when the horsepower had increased by 47%. On some of the earlier Holt-manufactured models, the steering clutches would freeze up in extreme cold. To regain steering there was no choice but to unhook implements and drive in a straight line to warm up*

Right and overleaf
1950 CHAMBERLAIN 40K. *This machine was manufactured by Chamberlain Industries, Welshpool, Western Australia. After the Second World War, there was a huge shortage of tractors, especially the heavier models. In 1947, Harry Chamberlain of the Australian Ball Bearing Co. in Melbourne, moved to Western Australia and with substantial Government assistance started the new tractor company in an old munitions factory. The Chamberlain proved a reliable tractor, even though a little heavy on fuel, and this example was in use for 28 years. The "Made in Australia" sign perhaps sums up the fifties era for Australians, finally starting to be fiercely proud of their own country, and not just another British colony. Home produced transport had also finally arrived with Australia's first car, the FJ Holden, produced in 1948, which coincided with the release of the Chamberlain 40K. The 40K had a two cylinder side-valve horizontally opposed engine of $6\frac{1}{8}$" bore and $6\frac{1}{4}$" stroke, giving 9 forward and 3 reverse gears, with speeds up to 18 mph*

CHAMBERLAIN
MADE IN AUSTRALIA

1922 CLETRAC MODEL W. *Made in Cleveland, Ohio, U.S.A., by The Cleveland Tractor Co., the Cletrac was fitted with a Weidley OHV engine, with a 4" $\times 5\frac{1}{2}$" bore and stroke producing 20 hp. An Eisemann impulse magneto was fitted, and the machine used $1\frac{1}{4} - 2$ gallons per hour from its 20 gallon fuel tank. It had one forward and one reverse gear driven through a spur gearbox. The Cletrac proved so successful that it remained virtually unchanged until 1931, and the above unit was in constant use for 45 years. In fact it is still used on Show Days for various towing jobs. Some models were used on saw mills until the 1960s, as the crankshaft pulley running at engine speed was a distinct advantage. In its time, it was able to plough 6-7 acres per day with a 5 furrow plough. Because of its compactness and reliability and the fact that it steered so easily it was a very manageable unit, unlike many of its competitors. One of the more embarrassing moments for one of the local Cletrac owners was when his tractor "threw a conrod" and he had to tow it home past his neighbours with a team of horses*

The Yorkshire Steam Wagon was built in Leeds, Yorkshire, England in 1903 by the Yorkshire Patent Steam Wagon Co. Fitted with a T type boiler, it produced 180 lbs working pressure of steam. One cwt of coal and 300 gallons of water were used for 30 miles of driving. The truck was capable of carrying 6 tons and pulling about 20 tons in two trailers towed behind. Originally used for carting wool, it is seen here with a typical load of wool and one of the new-fangled tractors with its fuel

1923 CLETRAC MODEL F. *This little crawler tractor was the smallest made by The Cleveland Tractor Co., Cleveland, Ohio. With its 3½" bore and 4½" stroke, it was rated as the 9-16 hp model. Unusually at the time, it was adapted to horse drawn equipment. It weighed 1820 lbs and sold for $595. Another interesting aspect is the high-drive sprocket, which the owner of this machine (Lynne Brandon) took up with the Caterpillar company. Caterpillar had brought out their new D7 in 1987, with its revolutionary new drive system, which helped keep the drive out of the mud. They were a little put out when it was pointed out that this had already been done 65 years earlier*

Above
1924 CLETRAC MODEL K. *The Model K is very similar to the 1922 model, but it boasted two forward gears instead of one. Here it can be seen devoid of all its panels and radiator. The four cylinder OHV Wisconsin motor can be clearly seen, along with the track runners with their 7 rollers rather than the earlier 3, which should have given a slightly smoother ride. The restoration by Merv Robinson is well advanced here, with the panels and the finishing off to be done – aspects which can often take just as long*

Above right
1953 COCKSHUTT. *This is an example of a tractor still in use 40 years on, seen here shredding pea straw for use as mulch for city gardeners. Made by Cockshutt Farm Equipment Ltd. in Brantford, Ontario, Canada, the owners purchased it because it was the most advanced tractor at the time. Their reasons for buying it were: its live power take-off (the power take-off was controlled separately so that it did not stop when the clutch was operated, as was usually the case); its large 15 × 34 tyres (normal size was 15 × 30); and remote hydraulics. The powerful 51 rated belt hp engine and 6 mph working gear proved a winning combination. The owner hitched two together to provide a 100 hp fwd tractor more than 10 years before the first one came on the market. The 6 cylinder Buda motor (3¾" bore and 4⅛" stroke) had the American Bosch fuel injection with a compression ratio of 14.3:1 and produced 46.22 drawbar hp at 1650 rpm. It had a range of gears from 1.6 to 11.6 mph*

Right
1908 DAIMLER RENARD. *This monster was made by Daimler in Coventry, England, in the form of a road train. Daimler acquired the manufacturing rights in 1907 from a Frenchman, M. Renard, who first demonstrated the "mechanised monster" in 1903. It had a 110 hp six cylinder Daimler sleeve-valve engine, which powered a unit described by the company as follows:- "It consists of a tractor, to which up to six trucks could be attached. Each truck [what we now call trailers] had six wheels, the centre pair being driven by chains through a series of propeller shafts running aft from the tractor. The front and rear wheels of the trucks steered in such a way that all the trucks followed a similar course." The vehicles were supplied to India, Canada, Australia, U.S.A., South America and Europe. Although they may have been successful elsewhere, this particular model had great difficulty in crossing dry river beds in the area it operated, which was largely devoid of any form of road. It did only six trips and then broke down irreparably and, together with its four trailers, it had to be towed to its destination, by a team of 60 donkeys*

50
DIESEL
COCKSHUTT
Grasslands

THE FARINA ROAD TRAIN
1908 RENARD-DAIMLER

1954 DAVID BROWN CROPMASTER. *The David Brown was manufactured in Melham, England, by David Brown Tractors Ltd. The four cylinder OHV engine produced 22 rated belt hp and 17 rated drawbar hp from its 3½" bore and 4" stroke engine at 1300 to 2000 rpm. Four forward gears were fitted giving a maximum speed of 12 mph and two reverse gears. The machine ran on kerosene and could pull 3,700 lbs. A two speed power take-off and belt pulley were available. The driving arrangement was somewhat unusual, as it was fitted with a dual seat, with the steering wheel in the middle so that the driver sat with the wheel to one side. Very awkward until you got used to it*

DEUTZ
DEUTZ

1953 DEUTZ FM2417. *The Deutz was manufactured by Klockner-Humboldt-Deutz, Cologne, Germany. It featured a two cylinder diesel compression ignition four stroke engine, of 4.7" bore and 6.7" stroke, producing 35 belt hp at 1350 rpm. Running on 12.75 × 28 tyres, it had five speeds up to 12.4 mph, with a reverse gear of 1.8 mph and weighed 6,600 lbs. This machine was found with the fuel pump rusted up after children had filled the diesel tank with water. Since repair, it has proved to be a very reliable unit and was used for broad acre farming. Although quite an advanced tractor in many ways, it is still necessary to remove the rocker cover and oil the tappets every 5 hours*

1920 EMMERSON-BRANTINGHAM. *Known locally as the "Everlasting- Bastard", they were noted for always breaking down, with badly designed transmissions and water pumps. They were also prone to tipping over sideways, due to the high centre of gravity and very narrow wheel-base. The machine was made by the Emerson-Brantingham Implement Co. of Rockford, Illinois, U.S.A. The 4¾" bore and 5" stroke side valve engine developed 23.2 rated belt hp and 13.16 rated drawbar hp at 900 rpm, running on kerosene. It had two forward speeds reaching all of 2.77 mph and weighed 4,400lbs*

Right

With the demise of steel wheels and the pneumatic tyre revolution came other problems. A steel wheel never got a puncture and the grips could easily be replaced when worn out. The early rubber tyres did not have the materials and quality of manufacture that we know today. They could be easily punctured, especially on newly cleared land where a large stick or stone might go straight through the tyre. This picture illustrates a 'permanent' repair of baling wire neatly stitching up a severe gash

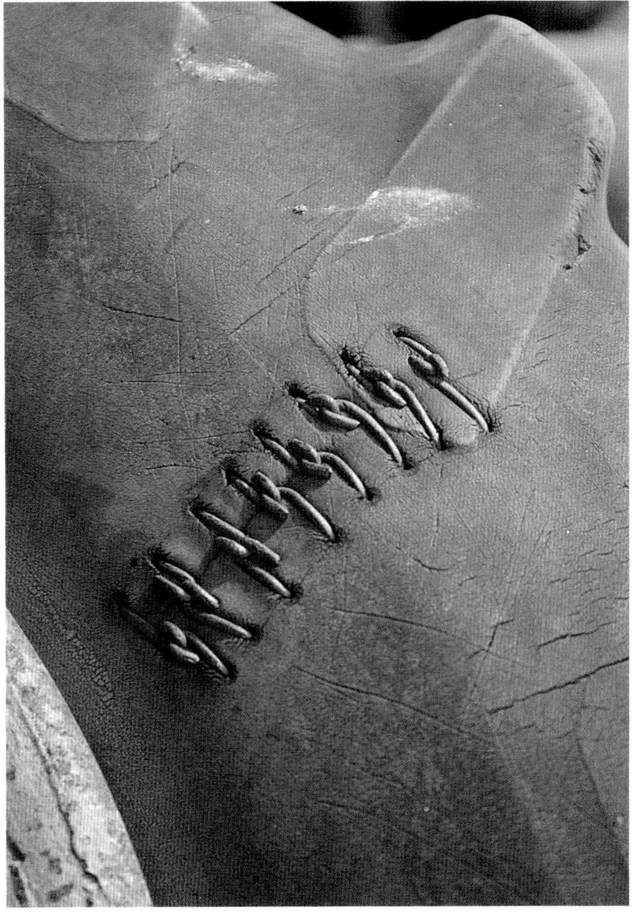

Left

Before the advent of large, wide rear tractor tyres and the fitting of duals to provide extra traction, various manufacturers offered these bolt-on grips, such as these Anders grips, to do the same job. They were only partially successful, however, duals and larger tyre sizes doing a better job

FIAT
Fiat

1924 FIAT 703. *The Fiat was made in Italy, and had a four cylinder side-valve motor of 4¼" × 6" bore and stroke, producing 19 drawbar hp and 30 belt hp at 900 rpm running on kerosene. There was a planetary gear final drive reduction built into the wheel hubs. Other unusual features were its non-detachable cylinder head and a clutch which operated by being pressed down to engage it, rather than releasing it. The belt pulley is also unusual in that the wheel drive can be disconnected, and the gearbox used to give three forward and one reverse gear on the belt pulley. In addition, the drawbar can be raised or lowered, as well as being adjustable from side to side*

1923 FORDSON CRAWLER. *The Fordson Crawler was not manufactured as a crawler. It began life as a standard Dearbourne wheeled model of 18-20 hp made in Michigan, U.S.A. The conversion kit was then fitted, to provide a full track vehicle for use on hilly and difficult terrain. Whilst the idea was probably sound, the practicality certainly was not. Having a worm-drive differential, the simple brake drums for steering were not very effective, and the motorwould almost stall on turning. Driving it could be a difficult and nerve-wracking exercise at first. The crawler is seen here pulling a dump rake, which was used to rake up the mown hay into heaps in preparation for baling or carting to a hay-stack*

Fordson

Above

Based in Dearborn, Michigan, Fordson tractors were also built in large numbers in the United Kingdom and Ireland, and they ended up being sent all over the world. From the first experimental model that appeared in 1916, nearly 750,000 were built until production ended in 1928. The engine was a four cylinder side-valve unit, developing 18-20 hp at 1000 rpm. It had a 4" bore and a 5" stroke. The tractor weighed 2,700 lbs and could pull about 2,400 lbs and featured three forward gears of 1.58, 2.24 and 7.05, with reverse. The Dearborn Fordsons left much to be desired, having vibrating coils as on the model T Ford, which caused difficult starting, and an unsusual carburettor, which provided more wind than fuel. Some farmers found it hard to believe that anything could get so hot in summer and still freeze in winter. The English Fordsons, produced until 1940, were a great improvement over the Dearborn models, having an increased bore size of 4⅛" with high tension magneto ignition and an efficient water pump. The Fordson is seen here pulling an early stone rake. The intention was that loose surface stones would be raked up into large rows and then loaded with stone rakes into tip drays

Right

This particular Dearborn example has special rear wheels made by the Tractor Grip Wheel Co., Toledo, Ohio – the idea being that they were self cleaning and did not block up as much as standard wheels with lugs

1948 FORDSON MAJOR HALF TRACK. *Made by the Ford Motor Co. in Dagenham, Essex, England, the Fordson Major Half Track had a side-valve four cylinder 4¼" bore by 5" stroke engine which gave 32.5 belt hp at 1200 rpm. It would use between 1½ – 2½ gallons of kerosene an hour. Three forward gears were fitted, driving the half-track arrangement, which seems to have been designed back to front, because when put into reverse, the tracks could run off the drive wheels and bend the back idler axle. The basic tractor, however, was pretty reliable. Geoff Nicholls is doing all the hard work. The tractor is pulling a contour plough, which is used on the steeper hillside to stop the soil from being washed away. The land is first surveyed to provide a line for the contour bank, which should fall 4" in 50 feet to provide a natural run-off in a heavy downpour. The plough then ploughs in both directions to create a bank to trap the water, and the land is then worked between the contour banks*

Fordson

Fordson
FORDSON

1924 GLASGOW. *One of the most unusual tractors produced was the Glasgow, a three-wheel drive machine built by Wallace (Glasgow) Ltd, in Cardonald, Glasgow, Scotland. The 27 hp side-valve motor had a bore of 4" and a stroke of 5½". This was the only one imported into South Australia, due to its failing in field trials against a 15-27 cross-engined Case. It remained unsold for several years, until one of Australia's 'cattle kings' bought if for some light belt work, and then left it to rot. After several years, it was resurrected and used for farming for a few more years. When found, the left hand front and the rear wheel crowns were broken. The last owner had a workman who, while driving it one day, hit a large pothole and the machine bucked him off. The Glasgow, careering out of control, hit a gum tree, and because of the large spikes it drove itself up the tree, until it stood on end, and then promptly fell over*

1927 12-24 HART PARR. *A typical wood sawing scene – wood being so many farmers' only source of heat during the winter. Also many tractors of this type ran saws in saw mills, some for several decades. The Hart Parr 12-24 model E was made by the Hart Parr Co., Charles City, Iowa, U.S.A. The 1926 Chevrolet Superior K Utility Express could carry one ton at 30 mph, with a 21.6 hp 4 cylinder OHV engine. From the radiator to the gearbox, it was identical to the car, only the tyre size, rear axle and chassis length were different. The Hart Parr had a two cylinder OHV engine of 5½" bore and 6½" stroke, developing 24 belt hp and 12 drawbar hp, using 1½-2 gallons an hour. It had 46" rear wheels and 28" front wheels. The machine was renowned for easy starting, (generally second pull), its reliability, as well as its frugal fuel economy, especially compared to other tractors of that era. The belt pulley for this model was an optional extra, but once fitted, proved to be a very useful addition. There was a very efficient fan and Madison Kipp force fed lubricators. The crankcase had overflow pipes, so that after lubricating the engine, excess oil channelled through to the final drive gears. The fuel tank held 12 gallons. First introduced in 1924, the model E was produced until 1928, when the model H was offered until 1930. The 'H' was the same except for a 5¾" bore rather than 5½".*

1927 28-50 HART PARR. *This was the largest tractor Hart Parr made in this series. The 28-50 was basically two 12-24 engines put together on a common crankcase, giving a four cylinder engine of $5\frac{3}{4}$" bore and $6\frac{1}{2}$" stroke. With the optional extension rims and the extra wide engine, it certainly makes an impressive sight from behind. Starting could be a little smokey, as one of the operations before starting was several turns on the Madison Kipp force feed lubricators to prime the engine with oil. Lubrication was a drip feed total loss system. The 50 rated belt hp was at 850 rpm. List price was $2,085 and about 1200 were made*

1937 OLIVER HART PARR.

Set against a typical turn-of-the-century colonial homestead, the 1937 Oliver was the first of the in-line four cylinder OHV engines. and the last of the Oliver line still to be called Hart Parr. The petrol-kerosene engine of 4½" bore by 5¼" stroke produced 28 rated belt hp and 18 rated drawbar hp, running at 1190 rpm. It was run on 12.75 × 24 rubber tyres and built in the Oliver Corporation's factory in Chicago. Power take off was standard. The Oliver had three speed transmission and weighed 3800 lbs. Subsequent models were simply known as 'Oliver'

1919 HOLT RENOWN CRAWLER. *This was made by the Holt Manufacturing Co., in Stockton, California. They built their first vehicle, a steam tracklaying tractor, in 1890. The first gasoline powered tractor appeared in 1908, after Benjamin Holt bought out Daniel Best. Holt became a leader in the tractor industry, with crawler tractors as his speciality, and when C.L. Best and Holt merged their interests in 1925, the Caterpillar tractor company was formed at Peoria, Illinois. The Renown had an engine with four separate cylinders, each with a 6" bore and 7" stroke, developing 45 hp (25 drawbar hp). It could pull 4,500 lbs on the drawbar and weighed 13,900 lbs. The rocker gear on its OHV engine was exposed and the engine alone weighed a hefty 2,300 lbs. Two forward gears were provided, giving up to 3½ mph and the engine was fitted with a Madison Kipp lubricator, a Kingston carburettor and a K-W magneto. Originally bought for farming, it was too expensive to run and too heavy on wet ground. The manufacturers made some amazing claims, and in one testimonial there was a glowing report about how wonderful the '45' was, but two years later, the same person sold it because it was too expensive to run. The testimonial said that the tractor used 112 gallons of fuel to plough 130 acres. It was then bought for use in pastoral country towing heavy equipment about. One of the advertisements enthused: "A short cut to work, with a 45 hp Caterpillar, through swamps impassable to men and horses." The reality was that on numerous occasions they were bogged down to the top of the tracks and had to be pulled out by a team of horses. On its last run in 1934, it used 20 gallons of fuel to cover 9 miles*

2 TON

1927 HOLT 2 TON. *This was one of the most popular tractors of the mid-1920's. Produced by the newly formed Caterpillar Tractor Co., it had a gear driven overhead camshaft and fan, with a 30 hp (15 drawbar hp) engine of 4" bore and 5½" stroke. Three forward gears and one reverse were fitted as standard, and the whole unit proved very reliable with the engine having a long service life. Much new country in the hilly areas was opened up due to its ability to climb the steep slopes. There is an interesting story told by one of the locals involving a 2 Ton. Two neighbours, who both had 2 Tons with land over 15 miles apart, were concerned about how long it was taking to travel the distances. One of them hit on the idea of transporting them to the location on a truck. This they duly did and being too lazy to make a proper ramp, they would simply drive the tractors off the back of the truck. Such a practice certainly did not do the truck a lot of good, because the front would rear up and then come crashing down as the tractor went off the back. It achieved the result for a while, however, until one day they forgot to remove the drawbar from the tractor, and as it left the truck, an almighty noise was heard as the drawbar demolished the tray of the truck. Needless to say, an unloading ramp was quickly built.The 2 Ton is seen here dam sinking, with a locally manufactured Linke-Noeh scoop. In South Australia, the land was originally divided into one mile square blocks, and in each block a dam was sunk in the line of one of the water courses. After rain, the dams would fill to provide water for stock during the long dry summer months. The tractor is driven by the owner, Hedley Koch. Ernie Masters controlling the scoop used to do the same job with a team of horses*

1958 HOLDER. *This sophisticated garden tractor had four wheel drive, articulated steering, power take-off and three point linkage. Made in Metzingen Wurtt, Germany by Gebruder Holder Maschinen Fabrik, it was fitted with a 12 maximum belt hp Sachs engine. The five forward speeds and three reverse allowed it to pull loads of 1,160 lbs on its 5.00 × 16 tyres. Seen here in its native orchard environment, it steadfastly refused to get bogged despite the wettest weather conditions on record. The Holder was restored and driven here by Roger Chapman*

1939 HOWARD MODEL DH22. *This little tractor was developed for use in orchards and market gardens. With its unique rotary hoe, it chopped up all remnants of the crop and put it back into the soil again. It was manufactured in Australia, by Howard Auto Cultivators Ltd., in Northmead, N.S.W. The 27 belt hp four cylinder OHV engine, of 3¾" bore and 4¾" stroke, ran at 1250 rpm, through a 5 speed transmission. The later models had 10 gears with the addition of a high-low ratio gear. It is seen here hoeing in the remnants of a brussel sprout crop. The following is an excerpt from an advertisement in 1947, under the heading, "All that is oldest is not best": "The spirit of Freedom, the concept of the Home in its best sense, Spiritual Values, and all the essentials which make for Stability and Peace seem naturally associated with bodies of men who, able to develop independence, and who through the nature of their daily tasks, contact with Nature and the love of Mother Earth, have the time and the environment to develop thought and lines of action which typify all that is best in a nation's life". So buy this machine*

1937 HSCS CRAWLER. *The HSCS Company started in 1842, with a Nathanial Clayton. Joining with Joseph Shuttleworth, initial production was of ship steam engines, which soon turned to agricultural machinery. Due to rapid expansion, European production was started in Vienna and Budapest, under a Mathias Hofherr, who then united with Mr. Schrantz, and the group combined to form the Hofherr-Schrantz-Clayton-Schuttleworth organization. They proceeded to make a complete range of agricultural machinery for world markets. The HSCS tractor was made in the Budapest factory in Hungary. It had a single cylinder crude oil two stroke steel-flap valve engine of 7½" bore and 7½" stroke, giving 25 belt hp and 20 drawbar hp. Ignition was by hot-bulb and it was fitted with three forward gears up to 2.75 mph and a reverse. When being restored, a new head was needed. One was duly found, but the atomiser was missing. The owner stratched his head and then proceeded to find the main part of the atomiser, in the back of an old blacksmith's shop, but part was still missing. He drove 5 miles into bush country to an old wrecked caravan full of baled hay. Moving about 40 bales of hay and then crawling under the hay like a rabbit, he found the tip of the nozzle, which measured 1½" by ½". What you might call 'finding a needle in a haystack'*

1952 HSCS STEEL HORSE. *This model had a single cylinder crude oil, hot-bulb, engine of 7½" bore and 9 7/16" stroke, producing 40 belt hp and 32 drawbar hp at 760 rpm. Starting was by a blowlamp placed on the head. The tractor was made to a high standard, and by 1950 there were over 50,000 in use worldwide. The noise and vibration proved beyond human endurance, however, and the previous owner said it had beaten most men who had driven it for any length of time. The last time he used it, the noise made him so ill he swore he would set fire to it – fortunately he did not*

1910 INTERNATIONAL MOGUL 10-20. *Built from 1909 to 1914, 2441 Mogul tractors were made at the Milwaukee Harvester Co. works in Wisconsin, U.S.A., by the International Harvester Co. The Mogul 10-20 resulted from the earlier development of the Morton traction truck to which International added their horizontal single cylinder four stroke petrol stationary engine. Developing 20 hp at 240 rpm from an 8¾" bore, it approaches what we now regard as the modern tractor.The engine was fitted to a frame and wheels with a rudimentary clutch and cooling system. It had rear wheels 70" high and 20" wide, with one forward gear of 1¾ mph and one reverse gear, and was capable of pulling two five furrow ploughs under the right conditions*

TITAN
10-20

1920 INTERNATIONAL TITAN 10-20. *During the period 1915 to 1922, over 78,000 International Titan 10-20 tractors were made by the International Harvester Co. of Chicago, Illinois, U.S.A. This made it one of the most successful tractors of the era, due to its reliability, simplicity and being a much more practical machine than the single cylinder Mogul 10-20 it replaced. The twin cylinder engine had a bore of 6½" and a stroke of 8", rating it at 20 hp at 575 rpm. There were two forward speeds of 2¼ and 2⅞ mph, with one reverse gear. This Titan is fitted with self-steering gear. Once the initial plough run had been done around the perimeter of a field, the self-steering gear would lead the tractor around the field. It led one of the local farmers to have many a hair-raising moment, as he would let loose the Titan to plough round a field and then follow behind it with a team of horses and a second plough. The only thing we did not find out was what happened on the corners, but at least it would have kept him very fit*

10 · 20
10 · 20

1924 MCCORMICK DEERING. *This was made by the International Harvester Co. and was very advanced for its time. It set the format for tractor design, which is still in use today, with the in-line four cylinder OHV engine (4¼" bore and 5" stroke) developing 20 hp at 1000 rpm.The one-piece cast iron frame ran from the radiator to the drawbar, and the engine, gearbox and differential were fitted in as one built-up unit. It had three speed transmission, up to 4.25 mph, and was one of the first tractors with power take-off for driving machinery. This McCormick Deering proved a very reliable unit and was in use for 25 years. The 10-20 was first produced in 1923 and continued to be made until 1939. During the later years, rubber tyres were available as an option*

Above

Appearing to be a normal farm shed on the outside, this unrestored collection represents several lifetime's work for someone with the money and ambition to tackle it. At least this is about 60 machines which are preserved indoors, rather than ending up on the scrap heap

Left

The South Australia embalming process at work. The dry climate and the sheer bulk of the tractor – which makes scrap salvaging a major headache, particularly when tyres have rotted away – have conspired to preserve many machines, some of which have become part of private collections or were taken on by the Booleroo Steam and Tractor Preservation Society. The hardiness of some of these machines is extraordinary: the 1924 Hart Parr on page 54, for example, was abandoned on an open hillside for 20 years. The carburettor was drained of water, the fuel and oil tanks filled, a few winds on the Madison Kipp drip feed engine oil lubricator – and on the second pull of the starting handle the engine burst into life

McCORMICK-DEERING
10-20 - H.P.

International 10-20's: 1910, 1920, 1924. This photograph shows the rapid development of tractor design over a period of 14 years. The 1910 Mogul is a large slow-revving (240 rpm) single cylinder, delivering its power at $1\frac{7}{8}$ mph, while the Titan has twin cylinders of $6\frac{1}{2} \times 8$", and revving to 575 rpm. The final evolution, from which the basic format has remained with little change to the present day, was the McCormick Deering. It had a four cylinder in-line engine built into a cast iron frame. The idea was to pull a smaller implement more quickly, so whereas the Mogul would pull two 5 furrow ploughs at $1\frac{7}{8}$ mph, the McCormick Deering would pull one twice as quickly, but on a tractor physically half the size

1927 MCCORMICK DEERING. *Formed from the great rivalries of two large grain harvesting companies, William Deering and McCormick, the International Harvester Co. was established on August 12th, 1902. There was talk of a merger nearly 20 years before, but the fierce competition had to take its toll before the merger was agreed. The 15-30 made history in 1921 when it arrived, with its unit construction and one piece cast iron frame. The 29.67 rated drawbar hp and 20 drawbar hp four cylinder OHV engine, with 4½" bore and 6" stroke, made it one of the larger tractors of its time. Said to replace between 8-10 horses, it could pull 4190 lbs at 2.39 mph, being fitted with three forward gears of 2, 3 and 4 mph. It became one of the most popular tractors of the steel-wheeled farming era, with many converted to rubber tyres. It is seen here using its belt pulley to drive a corn crusher, through which various types of cereal grain were put to provide flour for bread and to feed livestock. The truck is an early Model T one ton unit, loaded to its maximum of 12 bags of wheat*

McCORMICK-DEERING

1932 MCCORMICK DEERING W30.
The W30 was built in Chicago by the International Harvester Co., and produced 31 rated belt hp and 19 rated drawbar hp. The engine had four cylinders with overhead valves, 4¼" bore and 5" stroke and ran on kerosene. Three forward gears up to 3¾ mph and one reverse were provided. The W30 gave trouble free service over its 25 year working life.

1938 MCCORMICK DEERING F14 FARMALL. *The Farmall was a small rowcrop tractor, suitable for orchards and market gardens, and was first introduced in 1924 in the larger F20 version. The smaller F12 (the forerunner of the F14) was first made in 1932 and with its 3" bore and 4" stroke, it developed 15 rated belt hp and 11 drawbar hp at 1650 rpm. It had three forward gears from 2¼ to 3¼ mph. Production of the updated F14 by the International Harvester Co. only lasted for one year in 1938-9. The narrow wheels set so far apart give the tractor a stark, spindly and precarious appearance. The machine was restored by Brian Neumann*

F-14
McCORMICK-DEERING
FARMALL
INTERNATIONAL HARVESTER COMPANY

BE CAREFUL

Left

1947 FARMALL CUB. *The 'Cub' was the smallest tractor built by the giant International Harvester Co. First introduced in 1947, it had a four cylinder, four stroke engine, developing 8.3 rated belt hp and 6.6 rated drawbar hp from its 2.675" bore and 2.75" stroke. Revving to 1600 rpm, it could pull 1,596 lbs at 2 mph. Three speeds up to 6.13 mph were possible on its 8 × 24 rear tyres. This tractor proved immensely popular on small acreage farms, as well as on park lands and large lawns where it was fitted with underslung rotary mowers. The whole drive-train was built to one side to allow for increased vision in row crop work. Ideal for market gardens*

Above

1948 FARMALL A. *The 'A' was first announced in 1938, being manufactured until 1948 when it was replaced by the Super A. The 'A' had a four cylinder 3" bore and 4" stroke petrol engine producing 16 rated belt hp and 13 rated drawbar hp at 1400 rpm. With a four speed gearbox, the top speed was 10 mph. and it could pull a load of 2,387 lbs at 2 mph. Electric starting and lighting were fitted*

1948 FARMALL M. *The Farmall M was one of International Harvester's most successful tractors, 35,000 being built in 1951 alone. The 33.35 rated belt hp and 25.83 rated drawbar hp made it a very good all-round tractor. Its 3⅞" bore by 5¼" stroke four cylinder OHV engine, running at 1450 rpm, and through its five speed gearbox, speeds were 4.3 in fourth and 16 mph in top, on 11.25 × 36 tyres.The tractor is seen here hopelessly bogged, pulling a recently felled log. Wet weather, as much as dry, can cause no end of problems, and the scars from a badly bogged tractor can be seen for years to come. Four days previously, this particular spot had 3" of rain in 12 hours*

1916 JELBART. *The Jelbart was the epitome of the original concept of the 'tractor', which was the placing of a stationary engine on a framework in order to pull a plough by means other than horses. It was made in Ballarat, Victoria, Australia, by Jelbart Propy Ltd. The two stroke crankcase poppet valve engine produced 6-10 bhp at 625 rpm. It had a single cylinder and was fitted with a piston with two diameters. The compression end had a normal piston and rings, but the bottom end of the piston was of a much larger diameter and helped to move the air in the crankcase through the transfer port on the return stroke. The fuel was then mixed into the transfer port on its way to the cylinder. It was claimed that it helped to clear the unburnt fuel ready for the next firing. A governor controlled the air valve in the crankcase which shut the air flow off when it had reached its governed rpm; hence there was an unusual exhaust note, as firing was completely erratic. The Jelbart had a working speed of 1-10 mph, plus a reverse gear. The high road speed was most unusual, most tractors did not have such a speed until the 1950s. The open final drive made it difficult for the owner to hear himself think. To start the engine, it was primed on benzene, started on lighting kerosene and run on crude oil. Notorious for starting fires, this one was used on a training farm to train soldiers returning from the First World War. It was claimed that when hot, it would run on anything – one of the more bizarre fuels being mutton fat*

Agent
FALKINER ELECTRIC CO
103 WILLIAM ST. MELBOURNE.

SUPER
PLUME
ETHYL
WETMORE
GRINDER MIXERS
McLEOD
NEPTUNE DEPOT
Mobiloil
The Worlds Quality Oil
WE SELL
LIGHT of the AGE
KEROSENE
COR
GOODYEAR
RUSTON
OIL ENGINES
SIMPLE & RELIABLE.
LOW FUEL CONSUMPTION
GUARANTEED & MAINTAINED.
RUSTON & HORNSBY (AUSTRALIA) PTY. LTD.
SYDNEY. P.O.BOX 1569 E
MELBOURNE. P.O.BOX 4508.
BRISBANE. P.O.BOX 366.E
Cletrac

Above

1927 JOHN DEERE MODEL D 15-27. *Replacing the Model N 'Waterloo Boy' in 1924, the Model D was the first John Deere tractor. 25.8 rated belt hp and 16.8 rated drawbar hp were generated from its twin cylinder 6½" bore and 7" stroke, 800 rpm engine. Two forward gears wer fitted, giving 2.45 and 3.27 mph. On the earlier tractors, due to the poor fuels and incomplete combustion, de-cokes were necessary. A novel and inexpensive way of doing a de-coke was to pour hard grain rice into the carburettor. Apart from the wonderful smell of cooked rice, it did the engines no harm and completely removed all carbon deposits.The Model D is here using its ample belt pulley horse power to drive an International hand wire-tie baler. With three men, it was capable of tying 400 bales a day*

Left

Whereas for horses one can grow one's own fuel, tractors rely totally on outside suppliers. and the ever increasing cost of fuel has been a contributing factor in many farmers' business demise. Ironically in the 1930s, at least in the US, it wasn't a factor in the economic devastation of agriculture. When the bottom dropped out of the market, it dropped out of every market, including oil, and thousands of wells were capped to stabilise prices

Above

1947 JOHN DEERE ROW CROP. *This model had a twin cylinder four stroke forward facing flat engine of 23.5 belt hp and 20.6 drawbar hp. The bore was $4\frac{11}{16}$" with a stroke of $5\frac{1}{2}$" and it revved to 1250 rpm. Remaining in production from 1935 to 1952, it was fitted with six forward gears, up to 10 mph and a reverse of 2.5 mph. Weight was 4030 lbs on 10 × 38 pneumatic tyres. Several different models were produced; for example, the 'BO' was for orchard work and the 'BN' had a single front wheel*

Right

1937 JOHN DEERE MODEL D. *The John Deere Model D was made in Moline, Illinois, U.S.A. and was fitted with a twin cylinder OHV motor with a bore of $6\frac{3}{4}$" and a stroke of 7". It produced 37 hp at 900 rpm, running on kerosene. Three forward gears were fitted, giving up to 5 mph, with a reverse of 2 mph and it weighed 5,690 lbs. This particular tractor has been used for over 15,000 hours in 39 years, with only one new set of rings and several valve grinds. They were so reliable and successful that they were made with little change for 30 years.The John Deere is being driven by Ferg Innes and is crossing a fast flowing creek after heavy rains. Some 40 years ago, when the same creek was flooded, one of the local women was about to give birth and, due to complications, had to be rushed to hospital. As the tractor was the only vehicle that was able to cross the raging torrent, it was used to carry her across. The Publishers would like to thank Mr Innes for risking life, limb and machine for such an excellent photograph*

JOHN DEERE
HN DEERE

1949 JOHN DEERE CRAWLER MODEL 'MC'. *The John Deere Co. dates back to 1836 when John Deere, a trained blacksmith, built his first plough in Grand Detour, Illinois. By 1847, he was making 1000 ploughs a year and moved to Moline, Illinois, to take advantage of better transportation facilities. John Deere's first tractor experiments were in 1912, with a motor plough. It was not until the buying of the Waterloo Gasoline Engine Co, in 1918, however, that, overnight, John Deere became a tractor manufacturer with an already proven design. This same approach resulted in the 'MC' crawler. From 1940-47, the basic model BO chassis was shipped to the Lindeman factories at Yakima, Washington. Mounted on crawler tracks, the John Deere Lindeman Crawler emerged. Finally, the first all-John Deere crawler, the 'MC' was produced at Dubuque from 1949 to 1952. The machine had a four cycle two cylinder engine of 18.89 belt rated hp and 13.70 rated drawbar hp from a 4" bore and a 4" stroke, at 1650 rpm running on petrol. Fitted with four forward gears, it was popular for orchard work*

WATER
TEMPERATURE
AMPERES
JOHN DEERE
TRACTOR
MODEL
MC
SERIAL NO.
JOHN DEERE
QUALITY FARM
EQUIPMENT
BE CAREFUL
DRIVE TRACTOR AT SAFE SPEEDS. REDUCE SPEED WHEN TURNING OR APPLYING INDIVIDUAL BRAKES. DRIVE SLOWLY OVER ROUGH GROUND.

Left

1926 LANZ BULLDOG HR2. *Built in Mannheim, Germany, by Heinrich Lanz, the Bulldog had a single cylinder two stroke flap valve engine developing 30 hp, and running on crude oil. The engine had hot bulb ignition and was started by putting a blow lamp under the bulb in the head, to create a hot spot, which ignited the fuel. The engine was cranked over by a starting handle, which also doubled as the steering wheel. The gearbox was the sliding spur type, having four forward and four reverse gears, the latter being obtained by simply running the engine backwards. This feature led to many a hair-raising moment, especially at idle or low revolutions under load, when the engine might suddenly reverse itself. The tractor would run over what was behind, leading to numerous accidents.The HR2 was a particular problem in Australia's harsh summer climate, as it was hopper cooled with no cooling fan. It was alleged that a 44 gallon drum of water was needed at each end of the field to keep it cool and operational. In 1928, however, this was corrected with the HR4, when an efficient radiator and cooling fan were fitted. This turned the Bulldog series into one of the most successful European tractors, and with a few changes, they were produced until the mid 1950s. The HR4 also had a reverse gear fitted, making it much safer. However, the engine could still be started backwards, and there were a number of reports from farmers who, being unfamiliar with the tractor, claimed that it seemed slow in the forward gear and very fast in reverse*

Above

1936 LANZ BULLDOG CRAWLER. *Also made by Heinrich Lanz in Mannheim, the Bulldog Crawler featured a 44 hp crude oil engine, with 8.86" bore and 10¼" stroke, running at 630 rpm. It featured a plate clutch and a two range six speed transmission. There were relatively few made, and the one pictured has been made up of the remains of two collected from 800 miles away.*

Overleaf

Pulling a seven furrow plough in some of the more hilly country: how land became available with the advent of the crawler tractor. The low centre of gravity and the large surface area for traction created by the tracks made it possible. Descending steep hills, however, could cause the radiator to spill over. In grazing country, the ground could be sown to a mixture of oats and barley. This particular hill has been classed as unworkable by the Department of Agriculture 'because of erosion problems,' even though it has been worked like this for nearly 80 years

BULLDOG

1953 K.L. BULLDOG. *Made in Melbourne, Australia by Kelly & Lewis & Co., to the Lanz Model 'N' patterns of Heindrick Lanz of Mannheim, the K.L. Bulldog featured an 8.86" bore and a 10.24" stroke, two cycle hot bulb compression ignition engine. It produced 44 belt hp and 41 drawbar hp at 600 rpm, with crude oil fuel. The tractor was only made for a short while, as it was soon superseded by a full diesel engined version. The K.L. Bulldog had six forward gears from 2.7 to 13.8 mph and two reverse gears of 3.3 and 9.8 mph. It was fitted with 14 × 28 tyres, weighed 8,064 lbs and ran on 1½ gallons of fuel an hour. The Bulldogs were renowned for their vibration, the Kelly and Lewis being the worst example. Many farmers can remember going home at night with their bodies still rocking to the motion of the Bulldog. Because they were two strokes, they had a tendency to throw out a lot of red hot carbon sparks, so in the summertime they could be seen in the fields, with those large hats on the exhaust acting as spark arrestors. At times, some farmers would leave them running all night, gently rocking backwards and forwards (i.e. from forward to reverse) to keep them hot to stop the carbon from flaking off and causing sparks. This machine is one of the numerous Bulldogs restored by Norm McCallum. John Deere bought out Lanz Bulldog in 1962, in order to obtain their patented oil pump. They then closed the factory*

LOW
REVERSE

1926 LAWSON S6. *This tractor was made by the John Lawson Manufacturing Co., New Holstein, Wisconsin, U.S.A. Fitted with a four cylinder Wisconsin motor with a bore of 4⅛" and a stroke of 5¼", it produced 32 rated belt hp and 15 rated drawbar hp at 1200 rpm. It weighed 4,588 lbs and was provided with two forward gears of 2 and 3 mph and one reverse. Lawson built their first tractor in 1915, and produced tractors until 1935, when the company went bankrupt. They were famous for their quality products which ensured their survival against stiff opposition, but they could not withstand several crop failures and the great depression. A notable feature was the almost automobile-type driving position and cockpit*

LOYD

1949 LLOYD. *After the Second World War, there was a tremendous shortage of tractors and this machine was a not very successful attempt to fulfil the demand by Vivian LLoyd & Co., Camberley, Surrey, England. It was built up from a Turner marine diesel engine, coupled to a Ford truck gearbox and rear axle, and had Bren gun carrier tracks. As it had only Ford rear brake drums operated by levers, its steering ability was limited. Also the tracks were deficient, because the pins through the track plates kept breaking. They were simply not heavy enough for long term continuous use. The Lloyd's Mark I Turner V4 OHV diesel engine produced 33 belt hp and 25 drawbar hp at 1500 rpm with a 3¾" bore and a 4½" stroke. It was used mainly for dam clearing*

DIESEL

1952 FIELD MARSHALL SERIES 3A. *The Marshall name first started in the English town of Gainsborough in 1848. Marshall were noted for their steam engines, both portable and traction. Although around 1908-11 they had built a few motor traction engines, basically traction engines fitted with either a petrol start paraffin engine, or a two cylinder oil engine. In the late 1920s with American internal combustion engined tractors taking over the market, they decided to use the Lanz Bulldog as the inspirational basis for a new range of tractors. Built as a full diesel, it went through continuous changes until the 3A became the new improved model with extra gears, better fuel injectors, and a better piston design and cooling system. To distinguish it as a new model, it was painted in these bright colours, which certainly upset a few Marshall owners use to the 'correct colour' of green. The 40 belt hp single cylinder two stroke diesel engine of 6½" × 9" bore and stroke turned at 750 rpm and was linked to a six speed gearbox. Originally started with a cigarette paper or a gun cartridge hit with a hammer, the new models had the option of electric start and lighting, as well as 'Adrolic' hydraulic power lift. On one occasion, one of the new Marshalls with electric lighting was being driven at night by remote steering. It was a particularly black moonless night and the farmer had put the lights on the seed drill, from which he was steering the tractor by extension steering. On going through a large ditch in the field, the lights suddenly went off, and he stopped suddenly because the drawbar pin had come out, while the tractor continued off into the night. Thus followed a mad dash to try and follow the noise and turn off thc tractor. His wife saw him a little earlier that night, although he was not in the best of moods. The tractor illustrated was restored by Daniel Allen*

MASSEY-HARRIS
MASSEY-HARRIS

1937 MASSEY HARRIS MODEL 25 *The Massey Harris Co. was founded in 1891, by the merger of two companies – the Massey Manufacturing Co. of Toronto and A. Harris & Son of Brantford, Ontario, Canada. The Model 25 was the first to bear the Massey Harris name, as previous models were known as the 'Wallis'. These had been made by the J.I. Case plough works. Massey Harris had purchased this part of the Case group in 1928 in order to get into the tractor business effectively, their earlier forays into the business having failed miserably. The Model 25 had a four cylinder OHV engine of 4 ⅜" bore and 5 ¾" stroke, producing 41 rated belt hp and 26 drawbar hp at 1200 rpm, running on distillate or kerosene. There were three forward gears and one reverse. The present tractor needed major repairs, as it had run out of oil and the engine had seized. The gearbox housing had also been broken off, so another rear end had to be found to make the unit complete. The view from the rear shows one of the first power take-off (PTO) units fitted as standard to a tractor, which enabled many machines to be driven at a constant speed from the tractor, allowing machines such as harvesters to perform far more efficiently. This was preferable to the implement driving its own machinery from one of its wheels, which meant that the machinery stopped when the wheels stopped turning. There was also a water tank on this tractor, which fed small amounts of water into the inlet manifold to prevent pre-ignition in the engine, and to help it run cooler under heavy loads, especially when running on distillate fuel*

IMPERIAL OIL TRA
MAKERS MELBOURNE
Imperial

Left

1908 MCDONALD IMPERIAL. *The McDonald was made in Melbourne, Australia, with a 30 bhp two cylinder engine (6¼" bore and 8¼" stroke), and used 1½–2 gallons of fuel an hour. It weighed six tons and stood 9'8" high, 8'7" wide and was 13'6" long. In 1974, it out-pulled a 110 bhp John Deere tractor, although it did a 'wheelie' in the process. The owners comment was "It's all a matter of torque". The machine had a foot operated band brake on the live rear axle, with a wheel operated parking brake. It was fitted with a differential with a locking pin in the left hand wheel. The sales slogan at the time was: "It keeps your boys on the farm, because it reduces drudgery and makes farm work more interesting"*

Above

1938 MCDONALD IMPERIAL SUPER DIESEL.
The McDonald Model TWB Super Diesel was manufactured in 1938 by A.H. McDonald & Co., Ballarat, Victoria, Australia. Originally designed as a single cylinder stationary engine, it was modified to adapt it into a tractor frame with gears – very successfully, judging by the reliability reports. The two stroke steel flap crude oil engine of 9¼" bore and 10" stroke, providing 40 hp at 535-550 rpm. It was started by a blow lamp heating up a 'hot-bulb' to provide the initial ignition, and thereafter the combustion keeps the 'hot-bulb' glowing red to provide continuous ignition. This tractor has done in excess of 8,000 hours work with its three forward gears

The McDonald seen here with special 'More-Clay-Wheels' (to handle heavy clay soil) is pulling a six furrow plough. Initially it was not possible to see either the tractor or the plough for smoke when first put under load, as it had not worked for over 30 years. However, it soon cleaned out and was pulling the plough with ease. There were 806 McDonald Imperial TWB tractors made between 1931 and 1944 – this particular one being bought in unusual circumstances, as a stack of hay was traded in as part payment. A letter from McDonalds asked the owner how they could sell the hay and whether he had any contacts who might buy it

T55
Imperial

1920 MOLINE MODEL D. *The Moline was the result of the acquisition of the Universal Tractor Co., Columbus, Ohio, by the Moline Plow Co., Moline, made in 1915. The Moline had a four cylinder engine which replaced the two cylinder version in 1918. With a 3½" bore and a 5" stroke, it produced 22 rated belt hp and 12 rated drawbar hp at a high 1800 rpm. Fitted with an electric starter and lights, and with battery ignition, it was quite revolutionary. Other features were its turning brakes and differential lock, but it had only one forward and one reverse gear. The restorer, Merv Robinson, is at the controls. The Moline above is driving a chaff cutter, for which it was ideally suited. The chaff cutter chopped up the sheaves of hay into chaff or small pieces that could be more easily fed to and digested by farm animals. Concrete was added to the new model to lower the centre of gravity to stop the tendency for the machine to turn over. The rear wheels could be removed and replaced by various implements such as ploughs and cultivators. The two levers in front of the steering wheel are for the clutch and the accelerator*

MOLINE
ILLINOIS
MOLINE UNIVERSAL

Above
1949 NORMAG DIESEL. *This was the first shipment of German tractors to enter Australia after the war. They were manufactured in Zorge, Germany, under the joint control of the British and the United States occupation authorities. Fitted with 13.5 × 24 tyres, it had four forward gears and one reverse and weighed 3,800 lbs*

Right
1947 MM TWIN CITY MODEL UTS. *Dusk sets on the Minneapolis Moline UTS, the Minneapolis company being formed in 1929 by the merger of three smaller companies. The new model line-up of 1939 featured streamlined hoods, electric lights and fenders. The UTS featured an in-line four cylinder OHV engine with a 4¼" bore and 5" stroke, which produced 38 rated belt hp and 30 rated drawbar hp at 1272 rpm. Its fourth gear gave it a working speed of 6 mph and in fifth gear a very useful road speed of 20 mph*

Above
The International Mogul towers over the Peterborough and the Minneapolis Moline – twice the size and half the horsepower

Left
Oliver 90 Chicken Coop. This unrestored Oliver 90 has proved to be home from home. As it was tucked away in the shed, the productivity had gone unnoticed, hence some of the eggs were liable to be rather explosive

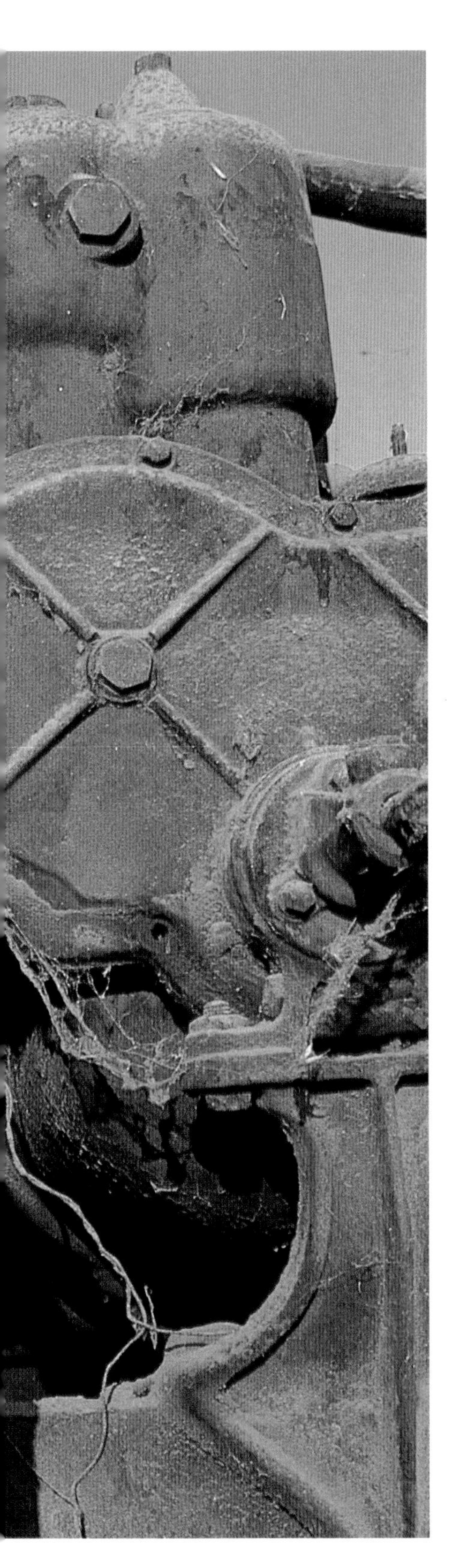

Above

1952 OTA. *The OTA came out of the Oak Tree Appliances Co., Coventry, England, with a four cylinder water-cooled Ford engine of 72 cu.in. developing 17 bhp at a maximum speed of 2000 rpm. The six speed gearbox gave a range of gears from 1 to 15 mph. 6.00×22 tyres were fitted to the rear and a 5.50×16 to the front. It was possible to adjust the rear wheels to be between 42 and 60 inches apart. Weight was 1,456 lbs and optional extras included a three point hydraulic lift and a four speed belt pulley. A four wheeled version was also available*

Left

1918 PARRETT. The Parrett originated in Chicago, Illinois, and had a 25 bhp (12 drawbar hp) Buda engine with a $4\frac{1}{4}$" bore and $5\frac{1}{2}$" stroke, running at 900 rpm. It had a cross-mounted engine driving through spur gears to the final drive, together with two forward speeds of $2\frac{1}{2}$ and 4 mph and one reverse. The rear wheels were 60" in diameter and 10" wide and the front wheels $46" \times 4"$. It weighed 5,000 lbs and sold at $1450 in 1916. Heavy use was made of SKF and Hyatt ball and roller bearings. This example was last used as a winch for land clearing, with the drive axles disconnected and a winch drum fitted to the drive axle

PETERBRO

Left

1922 PETERBOROUGH L30.4. *The 4¾" bore and 5½" stroke engine in the Peterborough tractor was said to have been originally developed as a tank engine for the British Ministry of Defence in the First World War, by Peter Brotherhood Ltd., in Peterborough, England. Running on kerosene, the four cylinder OHV engine produced 30-35 belt hp and 18 drawbar hp at 900-1000 rpm and was capable of running as low as 300 rpm on a Zenith carburettor and a Simms magneto. (It certainly wins most of the 'Go-Slow' races at the annual Steam and Traction event.) Maximum drawbar pull was 3,750 lbs at 2 mph, with a second speed available*

Above

Each piston had its top portion containing the rings working in the cylinder, while the piston skirt worked in an independent guide. This was claimed to prevent oil from the crankcase and fuel from the combustion process from mixing. The idea was to build an engine which would have lower friction losses, less mechanical noise (piston slap) and a smokeless exhaust. In fact, the engine managed to destroy its little ends, either through lack of lubrication or lack of metal around the gudgeon pin bosses. It also had gearbox main-shaft problems through a lack of central bearing support

Ransomes
RANSOMES
England
FRUIT GROWERS & MARKET
GARDENERS SOCIETY LTD
ADELAIDE
IMPORTANT
OIL BEARINGS OF ROLLERS
EVERY 5 HOURS

1953 RANSOME MG 6. *"Out working two horses, the Ransome is a mighty little tractor for all farm work. SO SMALL in size and weight. SO BIG in drawbar power, in profit earning efficiency, in application to farm tasks." Thus went the advertising for this little orchard and market garden tractor, made in England. The 8 hp, 3.42" bore and 3.98" stroke single cylinder engine was four stroke with side-valves. There was three speed transmission in both forward and reverse. Fitted with PTO and either mechanical or hydraulic lift for the implements, it was quite a popular unit*

Above
1920 RENAULT CRAWLER. *The Renault was made in France, being built along the lines of a World War I tank. It had a 24 hp engine running at 1000 rpm. The most unusual feature of this machine was its slanted radiator in front of the driver. The radiator was cooled by a turbine-type blade cast into the flywheel, the air being ducted through the radiator. One of its less endearing qualities was the three-way dipstick for the oil, and if the tap were turned the wrong way (the French instructions having not been fully understood) then the oil could drain without the operator realising it. This machine was found on the side of a dry river bed, half buried in silt, and had been subjected to at least three floods, according to the tide marks left on the machine.*

Right
1929 RONALDSON TIPPET. *The Ronaldson Tippet Company made this tractor in Ballarat, Victoria, Australia from 1924 to 1932, but it continued to be sold until 1938. The four cylinder American Wisconsin 20-36 hp engines were ex-WW1 stock, but had weak bottom ends and were noted for throwing conrods. Great pains were taken toadvertise the fact that the rest was made in Australia, with photographs in the brochures of the transmission, etc. The transmission featured their own 'Super Drive Epicycle Gear Transmission', with its two forward speeds and one reverse. One of the features was that by removing a cover it was possible to change two gears around and have a new set of gearbox ratios. With an extra set of different ratios available as an optional extra, there were then four sets of gears to choose from. Quoted as having a 60 hp clutch, an optional power take-off was also available. The brake worked on the inside of the belt pulley and was claimed to be very effective. Because they entered tractor manufacturing at a late stage, Ronaldson Tippet offered their machines on interest-free terms, which nearly caused the company to collapse*

Right
1920 STOEWER 3 S17. *The Stoewer was manufactured by StoewerWerke Ltd. and originated from Germany from 1919 to 1922. The prototypes were made during WW1, with just 200 being made in total and this is the only known survivor. As the German agricultural industry had not yet recovered from the war, it could not offer anything to match the English and American machines. Of nearly 70 prospective tractor manufacturers, only about 7 actually produced a tractor. The Model 3 S17 had a 7359 cc side-vlave engine of 5" bore and 6" stroke. Weighing 5 tons, it was a large tractor for its era, although its drawbar pull was a rather meagre 850 kg (1,870 lbs) for its size. It was claimed comfort was improved by a well sprung front axle*

Above
1950 TURNER 4V95 *(Yeoman of England) The Turner was sold as a general purpose medium tractor, with some confident advertising: "If you can tell genuine quality when you see it, you will pick the Turner – electric starting and lighting and it handles a 10-12 horse team load". It came as standard with a 540 rpm PTO and a belt pulley which ran at 1050 rpm. The machine was restored by Ron Piggott. The Turner was powered by a V4, four stroke, four cylinder marine diesel engine. It had overhead valves with a 3¾" bore and a 4½" stroke. Using compression ignition, it developed 36 maximum belt hp and 27 rated drawbar hp at 1500 rpm. Originating from the Turner Manufacturing Co., in Wolverhampton, England, it had four forward gears of 2, 3, 5 and 16 mph on 12.75 × 28 tyres, and was able to pull 4500 lbs at 2 mph. The Turner engine was also used in the Lloyd tractors*

In my beginning is my end? The 1927 John Deere Model D is dwarfed in both physical size and power by its grandchild. The Model D produced 25 rated belt hp and 16 rated drawbar hp, while the 1985 model 8650 generates 296 hp. Back in 1927, the limit was about 10-15 acres ploughed or 20-25 acres cutivated/seeded per eight-hour day. The 8650 pulling a 47 ft (14.3 m) wide rig can rip up, cultivate, or sow seed at 35 acres per hour. This machine has in fact worked over 700 acres in 24 hours. A fairly comfortable 24 hours at that, with air conditioning and stereo in the cab – a far cry from the exhausting battle with the elements and the land back in the twenties. What will another 60 years of development bring? Will there be such a thing as a tractor, or, indeed, anything that we would recognise as a farm? Perhaps the 8650 will be restored and preserved by the Booleroo Internal Combustion and Electric Engine Preservation Society